LIFE IN THE — U.S. — AIR FORCE

by Mo Barrett

PEBBLE
a capstone imprint

Published by Pebble, an imprint of Capstone
1710 Roe Crest Drive, North Mankato, Minnesota 56003
capstonepub.com

Copyright © 2025 by Capstone. All rights reserved. No part of this publication may be reproduced in whole or in part, or stored in a retrieval system, or transmitted in any form or by any means, electronic, mechanical, photocopying, recording, or otherwise, without written permission of the publisher.

Library of Congress Cataloging-in-Publication Data is available on the Library of Congress website.

ISBN: 9780756579890 (hardcover)
ISBN: 9780756580063 (paperback)
ISBN: 9780756579999 (eBook PDF)

Summary: Gives readers a peak into daily life for U.S. airmen.

Editorial Credits
Editor: Mandy Robbins; Designer: Heidi Thompson; Media Researcher: Jo Miller; Production Specialist: Tori Abraham

Image Credits
Shutterstock: Picksell, background (throughout); U.S. Air Force photo by Airman 1st Class Ashley Perdue, 15, Airman 1st Class Eboni Reece, 8, Airman 1st Class Eugene Oliver, Cover, (top middle), Airman 1st Class Greg Erwin, 11, Airman 1st Class John Nieves Camacho, 19, Airman 1st Class Michael Shoemaker, 12, Airman 1st Class Taryn Butler, 10, J.M. Eddins Jr., Cover, (bottom), Senior Airman Anna Nolte, 5, Senior Airman Brittany Y. Auld, 21, Senior Airman David Carbajal, Cover, (top left), Staff Sgt. Joshua J. Garcia, 18, Staff Sgt. Stephany Richards, 17, Tech. Sgt. Christopher Boitz, Cover, (top right), Tech. Sgt. Christopher Marasky, 13, Tech. Sgt. Michael Mason, 16, Tech. Sgt. Robert Cloys, 14, Tech. Sgt. Lauren Gleason, 7, U.S. Army photo by Capt. Anthony Deiss, 9,

The appearance of U.S. Department of Defense (DoD) visual information does not imply or constitute DoD endorsement.

Any additional websites and resources referenced in this book are not maintained, authorized, or sponsored by Capstone. All product and company names are trademarks™ or registered® trademarks of their respective holders.

TABLE OF CONTENTS

Look up High!. 4

Where Airmen Live. 6

What Airmen Wear 10

What Airmen Do. 14

Physical Fitness Test 20

    Glossary. 22

    Read More. 23

    Internet Sites. 23

    Index 24

    About the Author. 24

Words in **bold** appear in the glossary.

LOOK UP HIGH!

Look up! See those airplanes? They are part of the United States Air Force. The Air Force is a brave team. It protects America from high above the clouds!

Men and women in the Air Force are called airmen. They keep us safe from harm. Their special skills and tools help them solve problems.

WHERE AIRMEN LIVE

There are about 320,000 airmen in the Air Force. They work on **bases.**

Many airmen live on the base with their families too. Base housing can range from small apartments to large houses. Bases have stores, schools, gyms, and playgrounds.

Tinker Air Force Base, Oklahoma

Some airmen only work on the base. They live off the base and drive to work each day.

When airmen move to a different base, they meet new people. Some bases are in other countries. Airmen can learn about different **cultures**. They see new places.

Blackhorse Camp in Kabul, Afghanistan

WHAT AIRMEN WEAR

Airmen wear **uniforms**. This shows that they are a part of a team. The clothing also protects them. Most Air Force uniforms have a **camouflage** pattern. It makes them harder for enemies to spot.

An airman's uniform has a name tag and special badges. The badges show what **squadron** they belong to. They show what job an Airman does. Airmen also wear boots or special shoes. They protect their feet.

WHAT AIRMEN DO

Airmen work in every state. They work in different countries too. Special training teaches them how to protect people.

Airmen can have many jobs. They might be a truck driver, a doctor, or a **pilot**. Airmen can be mechanics or air traffic controllers too. They use their skills to help people.

Sometimes, airmen travel far away from home. They go on **deployments**. Some deployments are to help people who are fighting. Sometimes, airmen bring water, food, and supplies to people who need help. A deployment can last a few months to a year.

During deployments, airmen leave their families. While away from home, they might live in tents. They eat in a **chow hall**. After work, airmen write to or call family and friends. They may read or play games to pass the time.

The Air Force guards the United States from enemies. Airmen are proud of their jobs. They chant "Fly, fight, win!"

PHYSICAL FITNESS TEST

Airmen have to stay healthy to do their jobs. They have to take an exercise test every year. Make up your own physical fitness test. You could try these or make up your own exercises.

- How many jumping jacks can you do in 30 seconds?

- How many sit-ups can you do in 30 seconds?

- How many push-ups can you do in 30 seconds?

GLOSSARY

base (BAYS)—an area run by the military where people serving in the military live and military supplies are stored

camouflage (KA-muh-flahzh)—special patterns or colors that help things or people blend in or stay hidden for safety

chow hall (CHOW HAHL)—where Air Force members eat their meals when they're on a base or during a deployment

culture (KUHL-chuhr)—a people's way of life, ideas, art, customs, and traditions

deployment (di-PLOY-ment)—when troops move to a particular location to prepare for military action

pilot (PYE-luht)—a person who operates a flying vehicle

squadron (SKWAHD-ruhn)—a smaller unit or group in the military

uniform (YOU-nuh-form)—special clothing that members of a particular group wear

READ MORE

Besel, Jennifer M. *U.S. Air Force*. Mankato, MN: Black Rabbit Books, 2023.

Morey, Allan. *U.S. Air Force*. Minneapolis: Jump!, 2021.

Vonder Brink, Tracy. *The United States Air Force*. North Mankato, MN: Pebble, 2021.

INTERNET SITES

Air Force
kids.britannica.com/kids/article/air-force/352717

National Museum of the United States Air Force
nationalmuseum.af.mil

United States Air Force Facts for Kids
kids.kiddle.co/United_States_Air_Force

INDEX

airplanes, 4

badges, 12

bases, 6, 7, 8, 9

chow halls, 18

deployments, 16, 18

exercising, 20

jobs, 12, 14, 19, 20

squadrons, 12

training, 14

uniforms, 10, 12

ABOUT THE AUTHOR

Mo Barrett is a retired Colonel. She spent nearly 30 years in the Air Force flying as a pilot and setting up airfields. She is now a public speaker and entertainer, using humor to change the way people laugh, learn, and think.